Intermittent Fasting

The Secret to Lose Weight, Slow Aging and Detox Your Body.Live an Healty Life.

By Megan Amber Stevens

Disclaimer

Table of Content

Table of Content   4

Chapter 1   10

   Why is it beneficial for you to manage when you are eating?   10

   Why does it work?   12

   What are the Benefits of Intermittent Fasting?   13

   How It Can Affect Your hormones and cells   16

   Combining intermittent fasting with the keto diet   17

   Who Should Avoid or be Careful?   17

     Can Women Fast?   18

   Side effects and Safety   19

Chapter 2   21

   1. The alternate-day method   22

   2. Time-restriction method   22

   3. Twice-a-week method   23

   4. 24-hour method   23

   5. The warrior day method   24

   6. Meal skipping method   24

   7. Fat-burn fasting   24

   8. Bone fasting   25

   9. Juice fasting method   25

Chapter 3   29

   2. Beneficial to heart   30

   3. Healthy Brain   30

   4. Reduction of cancer   31

   5. Boosting of biological processes   32

Chapter 4   37

Chapter 5   42

1.	The Easy Italian omelette (melt)	42
2.	Low Carb California Omelette	44
3.	Low Carb Veggie Bowl	46
4.	Low Carb Salmon and Cheese	48
5.	Low Carb Caprese Omelette	51
6.	Sausage and Guac Stacks	52
7.	Kale Eggs and Sausage Crumb	54
8.	Fat-Burning Vanilla Smoothie	56
9.	Spinach & Feta Omelette	57
10.	Blackberry Cheesecake Smoothie	59

Chapter 6   60

1.	Mediterranean Low-Carb risotto	60
2.	Jalapeno Popper Stuff-Burger	62
3.	Chorizo-Stuffed Spaghetti Squash	64
4.	Creamy-Pesto Tuna	66
5.	Beef Hot Pockets	68
6.	Pork-Chops with Asparagus and Hollandaise	70
7.	Caprese-Chicken bowl	72
8.	Portobello-Mushroom Mini Pizza	74
9.	Salmon Patties	75
10.	Almond-Maca Smoothie	77

Chapter 7   78

1.	Salmon-Asparagus &Blender Hollandaise	78
2.	Butter Braised Cabbage with Crispy Ham	80
3.	Salmon & Avocado Omelette-Wrap	81
4.	Bun-less BBQ Guac Burger	83
5.	Crispy Lemon &Thyme Chicken	85
6.	BBQ Meatball Skewers	87
7.	Italian Meatza	89

8.   **Salmon with Creamy Spinach Hollandaise**   91

9.   **Lamb Souvlaki**   93

10.   **Chocolate-Coconut Smoothie**   95

Chapter 8   96

Is it Ok to Exercise While fasting?   98

Making your exercise more effective while fasting   99

1.   Select the time wisely   99

2.   **Think about your macros to select your workout routine**   100

3.   **Choose appropriate post-workout meals to build or maintain muscles**   101

How can you do your exercise while fasting safely?   101

•   **Eat before your moderate- to high-intensity workout**   102

•   **Stay hydrated**   102

•   **Maintain your electrolytes**   102

•   **Adjust the duration and intensity wisely**   103

•   **Select the type of fast appropriately**   103

•   **Make your body the priority**   103

Is it difficult to exercise while intermittent fasting?   104

•   **Cardio Exercises While Intermittent Fasting**   105

•   **Intermittent Fasting and High-Intensity Interval Workout**   105

Is weightlifting advisable While Fasting?   106

Best options for exercise during Intermittent Fasting: 107

Workouts to must not do in a fasted state: 108

Plan for Working out While Fasting   108

Summary   110

Chapter 9   112

Why do people like to combine keto with Intermittent Fasting for Weight loss?   113

Do keto and intermittent fasting work better together?   113

Which people can choose an Intermittent Fasting Keto Approach?   114

The Correct Method to Start an Intermittent Fasting Keto Combination Diet   115

A Model Menu for the combination of Keto and Intermittent Fasting   116

  Day 1   116

  Day 2   117

  Day 3   117

The Potential Health Benefits of a Keto Intermittent Fasting Diet   118

Is combining Keto and Intermittent Fasting risky for your health?   118

The final words on Combining the Keto Diet with Intermittent Fasting   119

Introduction

As we hear with the passage of time, that intermittent fasting Is playing a pivotal role in helping our daily routines get settled. Since we are lazy enough to ruin our bodies, we want to make ourselves better in the minimum time and with the fastest way possible.

Hence, in this book, everything is discussed in detail. What intermittent fasting is? The benefits you acquire when you follow this procedure. How it actually works? And what you need to do in order to set a meal plan and diet plan? How many types does intermittent fasting have? Its side effects and things needed to be avoided, especially by women.

The importance of intermittent fasting in your daily life? And, what might happen to you if you don't follow it? Then you need to know the recipes you can follow to burn all that fat with your meal plan. All of them are discussed that are beneficial with time aspects: breakfast, lunch, or dinner. Since most of you don't like t gym and are very lazy to work out every day. The best possible solution to follow is intermittent fasting.

The diseases that can cling to your body can destroy your life. Their life expectancy rate can also be endangered. Women are also persistent in consuming fatty foods full of cholesterol, high fructose levels, high sugar levels, syrups, and trans fats even.

Hence you should look after yourself rather than wasting yourself. Dieting is even more difficult when you don't eat anything and simply eat salads. That's worse for your health since you do not intake proteins. You need to intake essential requirements that are needed by your body.

Since we really need intermittent fasting in our lives, it is best for everyone to follow accordingly. Even if you wish to burn that fat away.

Chapter 1

The fundamentals of intermittent fasting

Intermittent fasting is an eating pattern, not a diet. It is the process of setting up your meal plan in a way that makes you eat more. It does not affect what to eat, but rather when to eat.

Why is it beneficial for you to manage when you are eating?

Well, it is a good thing that you become slim without losing your calories or going with a diet plan. Indeed, you will try to balance your calories when starting intermittent fasting. Moreover, intermittent fasting helps you to maintain calories while getting slim.

All of that, most people go for intermittent fasting to lose body fat. Here we will discuss how a person loses body fat through intermittent fasting.

It is important to note here that, intermittent fasting is the best approach for losing the bad mass off while maintaining the good mass on, but it needs little change in behavioral pattern. Intermittent fasting is the simple approach that you can easily practice and get significant consequences from it.

How does it work?

To know the working of intermittent fasting, it is necessary to recognize the variations between the eating state and fasting state.

When your body is taking in and digesting the food, it means your body is in a state of feeding. Normally, the fed state begins when a person starts eating and its end after 3 to 5 hours when a body breaks down and absorbs the food. When a person is in a fed state, it is nearly impossible for the body to dissolve the fat because the body insulin level is very high at that time.

After this, your body is in state of post-absorption, a stage when your body is not digesting and absorbing the food, meaning that the body is at rest stage. This stage lasts for 8 to 12 hours, and the body enters in the fasting state. At this stage, your body loses fat because insulin level is low.

At this stage, your body burns fat more quickly than at feeding state. As your body cannot enter the fasting state after 12 hours of your last meal; therefore, there is less chance of burning body fat. This is the main reason that person lose fat more quickly when they start intermittent fasting without changing their eating patterns. Therefore, fasting state burns the body fat that you hardly achieve during the fed state.

Why does it work?

Your body reacts to the consumption of energy (eating the food) with the production of insulin. The higher is the sensitivity of your body towards the insulin. The more is the probability that you would get to use food which you consume effectively. Your body gains sensitivity towards insulin while you are fasting. Such changes related to insulin sensitivity and production can help in leading towards muscular creation and weight loss.

Two main things can be related to this:

A meal after the workout can be used much efficiently: As the food you consume gets converted into glycogen and get stored up into the muscles or get burnt in the form of energy instantly. And this helps you in the process of recovery, with minimum amounts that are stored in the form of fats.

Lesser fat gain compared to your regular days (without intermittent fasting): Having insulin sensitivity at a normal level, the food and carbs you consume will lead to full glycogen storage, resulting in sufficient glucose within your bloodstream. Thus, more likely, they will store in the form of fats. And this means to have fewer fat gains of your body.

For most of the various reasons based on physiology, intermittent fasting has proved to help in promoting weight loss by lesser fat storage and muscular building if carried out properly. In your body, the level of growth hormones gets raised while fasting conditions (both while you are asleep and after the duration of fasting). Along with this raised secretion of growth hormone, the decline in production of insulin (and so, raise in the sensitivity of insulin resultantly), and you are basically priming the body for muscular growth as well as a fat loss with the help of intermittent fasting.

A little less science-based version about intermittent fasting is that it teaches your body to make use of the food which you consume most effectively. Your body gets to learn how to burn the fats like the fuel as you are depriving it of the new calories for continuously pulling from (when you eat the whole day)

What are the Benefits of Intermittent Fasting?

There are many benefits of intermittent fasting. One of them is the fat loss of the body. Here are some other benefits:

Intermittent fasting can make your day simple.

People like to have behavioral changes, gain simplicity, and eliminate stress. You really enjoy the simplicity which intermittent fasting adds into your life. As you wake up, you never worry regarding your breakfast. You just catch a glass of milk and begin your day.

You enjoy eating, and you never mind cooking. Thus, there remains no more hassle of cooking three meals per day for you. But the intermittent fasting lets you have one lesser meal, and this means that to plan one lesser meal, to cook one lesser meal and to stress regarding one lesser meal in a day. It can make your life a little simpler, and many people like it.

Intermittent fasting aids you to live longer.

According to the latest research, it is found that one of the great ways to lengthen your life is to restrict the calories' intake. This makes a lot of sense in terms of logical reasoning. As you starve, your body will find a way to extend life. However, there can be only one issue: would you like to starve yourself just for the sake of lengthening your life?

Most of the people won't **like to do so, but they would be interested in having to fun their life as well as lengthening their lives. Starving themselves don't sound much appealing.**

Here is the good news: intermittent fasting works in the same way as calories restriction in terms of extending the life period. Or, you can say that you can enjoy many advantages of long life without starving yourself.

Back in the 1900s, it was found that the intermittent fasting has extended the life period of mice. And recently, the study discovered that an alternate day intermittent fasting could lead to a longer life.

Intermittent fasting reduces the risk of having cancer.

This point has remained a target of debate because of the fact that there has been no loads of experimentation and research done on the relation between fasting and cancer. However, early reports seem to be positive.

A study done on 10 patients of cancer suggest that the side effects related to the chemotherapy may get diminished by using the method of intermittent fasting before treatment starts. This research has also supported by another study carried out for the alternate-day fasting on the patients of cancer. The conclusion which was made that the fasting is done before chemotherapy can cause better results with high cure rates as well as fewer deaths.

In the end, a comprehensive analysis done of different studies on cancer and intermittent fasting has shown that this method cannot just diminish the risks of cancer, but the cardiovascular diseases too.

Intermittent fasting is a lot easier as compared to dieting.

A reason that most of the diets fail isn't that you switch to a wrong form of food, but it is because you don't follow this diet in actual over a long period. This isn't a problem related to nutrition. Rather this is a problem related to the behavioral change.

It is the point where intermittent fasting comes in as it seems to be very easy to imply after you get over an idea that you require to get fed all the time. For instance, a study found that intermittent

15

fasting seemed to be an efficient strategy for weight loss in obese adults. So, it was concluded that the subjects adopted the routine of intermittent fasting quite steadily.

How It Can Affect Your hormones and cells

As you fast, few things start to happen inside the body on the molecular and cellular levels. For instance, your body settles its hormone levels for making the stored fats easily accessible. Also, the cells begin the vital repair processes and alter the formation of genes.

Following are some changes which happen in the body during the fasting:

Insulin: the sensitivity for insulin get enhanced, and the amount of insulin decreases dramatically. The low levels of insulin can make the stored fats of the body easily accessible for you.

Human growth hormone: the level of the growth hormone increases drastically in your body, rising as high as 3-fold. It has advantages for muscular gain and fat loss, to mention a few.

Cellular repair: while fasting, the cells begin the cellular repair procedures. It includes autophagy, in which cells get digested and eliminate the dysfunctional and old protein, which made up inside the cells.

Gene expression: there can be some changes within the functioning of genes about the protection from diseases and longevity. Such changes in the hormone level, gene expression, and cellular functioning are responsible for the benefits of health during intermittent fasting.

Combining intermittent fasting with the keto diet

By having protein and fats only, your body should adapt to work on the fats as fuel rather than carbs. If there are no glucose/carbs, the body turns fats into ketoses and then uses them as a fuel, the process is known as ketosis. There exist 2 ways to do this process:

- To eat in a manner which induces ketoses (high fat, low carb)

- Fasting – that's what you have been reading!

Probably, you won't be "keto-addicted" (your body start running on the ketones) only by skipping a meal in one day – your body will have sufficient glucose still stored from the carb-based meals you take in dinner and lunch.

For using the ketosis along with intermittent fasting, your fats require to be quite long for depleting your glucose/carb storage, or you require to restrict carbohydrates strictly from the meals along with IF for entering into ketosis.

Who Should Avoid or be Careful?

Certainly, intermittent fasting isn't for everybody. If you have a family history related to an eating disorder or if you are underweight, you must not fast without getting a consultation from your physician first. In such cases, the process can be harmful downright.

Can Women Fast?

There are few pieces of evidence that the intermittent fasting probably doesn't result to be as advantageous for females as it can be for males. For instance, a study has shown that it enhanced the insulin sensitivity in males but worsened the blood sugar control in females.

Though there are not a lot of studies done on this subject on humans, studies on rats have shown that the intermittent fasting can make the female rats masculinized, emaciated, cause them to have their cycles missed and infertile.

There exist a great number of the anecdotal studies of females whose menstrual cycles have stopped as they began the IF and had to go back towards normal diet as they resumed their older eating fashion. For such reasons, females must be careful while intermittent fasting.

They must follow a separate set of guidelines, like to ease into the stopping and practicing instantly if they face any issue like amenorrhea (lack of menstruation). For the issues related to fertility or/and if you are trying to have babies, think of holding off on the fasting at the moment. This eating fashion is not probably also a good idea if you are breastfeeding or pregnant.

Side effects and Safety

One of the top side effects of intermittent fasting is hunger. Along with this, you may feel weakness. Your brain mayn't do as good as you are used to. Maybe this happens only for some time, as it may take some time for the body to get adopted to a new meal plan.

If you have any sort of medical condition, you must take advice from your doctor before you try intermittent fasting. It is especially vital if you:

- Are breastfeeding or pregnant

- Have diabetes

- Take medications

- Have low or high blood pressure

- Are underweight.

- Are a female who is trying to have a baby

- Have a family history of any eating disorder.

- Are a female with a family history of amenorrhea.

Having said that, IF is an excellent safety profile. There exists nothing harmful regarding not eating for some time while you are well-nourished and healthy as a whole.

Diets are quite easy in contemplation, hard in the implication. Talking about the intermittent fasting, it is completely opposite – it is hard in contemplation; however, easier in the implication. Many of you have contemplated while doing the diet. When you find a diet which appeals to you, it appears as if it will come like a breeze for you. However, when you get into this nitty-gritty, it turns out to be tough. For instance, you must stay on the low-carb diet nearly every time. However, if you are thinking to go on the low-fat diet, it seems easy. You think about the bagels, jelly, corn, mashed potatoes, whole wheat bread, bananas, etc. – each of these sounds attractive. But you were to embark on this low-fat diet. You would get tired of it soon and wish that you could have eggs and meat. Thus, a diet is easier in contemplation, however not very easy in a long-term implication.

Intermittent fasting seems hard on the contemplation; about that, there exists no doubt. People would start to ask "you go without eating for one day!", incredulously what you explain about what you are doing. However, once you have started, it is like a snap. Never worry regarding where and what to eat 1 out of 3 or 2 out of 3 meals of a day. It is an excellent liberation which you get. Your expenditures for food plummet. And you don't remain particularly hungry. Although it is hard to overcome the concept of being without food, as you start the regimen, nothing would be easier.

So overall, ease of the intermittent fasting can be considered one of the best reason for giving it a try. This offers a great range of health-related benefits without needing an extensive lifestyle change.

Chapter 2

Types of intermittent fasting

Everyone wants to have a proper body, just those perfect model looks and if not that then surely everyone wants to look their best. As beneficial as exercising can be, it is still difficult for some to keep doing it daily. Although, just by the phrase "no pain, no gain" doesn't mean there aren't different ways to be in shape. Just like dieting, there's intermittent fasting.

Intermittent fasting does not mean you keep a fast and don't eat or drink until your fasting time ends. It simply tells you when to eat and when not to eat. You don't reduce your intake; in fact, you don't reduce how much you eat at all. You change your lifestyles and diet timings to stay fit and healthy. There is a process involved in making a plan which you can follow for "Intermittent Fasting (IF)," and there are many types of intermittent fasting from which you can choose what you require.

This means that if you want to eat chocolates, pastries, pizzas, no one is going to stop you. You can eat anything you want. You need to know when you can eat your yummy dishes. You don't have to wait for months. That does not affect your body if you follow these steps of intermittent fasting. There are many more ways you might hear from other sources, but they aren't much tested or authentic.

Stay sure and safe of what you decide after thorough research because this is a matter of your health and body which should never be compromised. Here are some 9 popular and most effective steps mentioned:

1. The alternate-day method

This is simply the "modified" fasting technique. On one day if you are taking 500 calories or 25% of your normal diet then the other day, you will continue with your regular diet. (however, some of the days you'll need to go with 0 calorie intake.)

So, if you're taking a day off from calories, that won't hurt when you get the other day to eat anything of your desire. You don't need extensive workout. You just need to manage your timings.

2. Time-restriction method

This is the 16/8 or 14/10 method which means you don't eat 16 hours a day and only eat 8 hours of the day. This method is very popular because no one eats when they are sleeping, and this can be adopted whenever you like even once or twice a week. The same goes for the 14/10 method.

16/8: eating timings lie between 11 am and 7 pm or 12(noon) and 8 pm

14/10: eating timings lie between 10 am and 8 pm

Some people just increase their sleep timings or keep themselves busy with work or movies. It's all up to you how you want your schedule to be set. This method is proven and tested. This method is considered best because timely eating is what everybody needs, not overeating.

3. Twice-a-week method

This method focuses on 500 calories two days a week, with one meal of 300 calories and the other meal of 200 calories (it's better to take high fiber and high protein meals to keep you full.) you can choose your own choice of days like either it's Monday or Friday. However, you need to eat your normal diet on the remaining 5 days.

You don't need to overdo it. Just be cautious for two days and enjoy the remaining weak with the love for your food.

4. 24-hour method

This method includes not eating anything for straight 24 hours. However, it's mostly done once a week. For precautions, you need to eat your regular tummy-filling diets in the remaining 6 days.

Be careful, though. If you want to keep a 24-hour method of fasting from food, then you need to take those proteins and calories the day before it. You do not want a headache. So, take foods rich with proteins and beneficial ingredients.

5. The warrior day method

This method includes those days of the weak, in which you eat a very little amount of fruits/vegetables in the day (just to keep you stable i.e., example: one apple) and then having a full huge meal at night.

Some people do it three times a week for fast effects. However, you need to keep your night meals enriched with all those necessary proteins, vitamins, and calories. It is essential, so you get headaches.

6. Meal skipping method

You don't need to have a full tummy all the time. If you don't feel like eating and have enough energy, then you simply skip a meal. This method can be done whenever you want, even daily. Example: Having a full-fledged breakfast and then don't feel like having lunch? Maybe no meal at night and just a glass of milk will do the job? (Make sure to keep your protein intakes for no headaches.)

7. Fat-burn fasting

If you're really fat and weigh a lot and want to burn your fat away as quick as possible, then this method will help you. This method consists of the process of autophagy. Autophagy cleans the cells with new healthy cells. When it combines with ketosis, which means that when your body is low on carbohydrates, and your metabolism starts, then it burns your fat away with a bit of pain. You just need to fast and get yourself a ketogenic diet. It will burn your fat away.

During this process, you can have a specific coffee, which will boost this method and prove helpful to you. You can have up to 2tsp of MCT oil/butter, carb or protein food like coconut milk, cream, etc. will do the job.

8. Bone fasting

This method includes a bone broth, and you drink bone broth when you're fasting for an extensive time. People mostly opt for 24 hours. Some might even go for 36 hours, though that might not be possible for everyone.

When you are determined to burn away all that fat as quick as possible, and you go for a long fast and then suddenly you feel so hungry that you can't control, then take a bone broth and drink. This will not affect your calorie intake and stuff your tummy as well. If you eat anything else, that, of course, will have more calories. Having more calories might upset you if you want to be hasty in your slimming goal.

9. Juice fasting method

It's better to not opt for this method because the ketosis method does not apply here. However, this method proved satisfactory and beneficial for a lot of people. It is up to you in the end. The ketosis process must happen in your long fasting time. By drinking a glass of juice will give your body those carbohydrates and fructose. That will only interfere with your efforts.

If you want to lose weight, then you can lose a lot of weight with this method. Though, only some muscle and lean tissue will be burnt, not all fat. Hence, this method is not very smooth, to begin with.

In this method, you'd drink juices in your fast. You don't eat anything for the whole 24 hours and simply drink a glass of juice when you get hungry. This is known as juice fasting.

When we are talking about intermittent fasting, we need to be careful of what our body requires. And our body needs water at most. So, it's best to stay hydrated as much as it is possible throughout the day. You should drink loads of water, herbal tea, and foods that contain water as well. And of course, when we say you can eat anything you desire, try to keep the calories in a stable amount when you have just started this procedure. It also requires your keen analysis of your schedule setting.

When you wake up in the morning, the first thing everyone must do is, drink lots and lots of water. At least as much as you can, and this must be done daily. It is also called hydro-therapy. You take your breakfast at least 20 or 30 minutes after you drink water (the amount you can).

On your fasting days, it's best that you keep yourself distracted in other stuff and keep your mind away from food as much as you can. It's even better if you do yoga and relaxes your body and mind. That is if you want to. Try your best to avoid strenuous activities as it would be hard on your body and relax as much as you can. Do light work only.

Tip: if you get agitated with your fast and its starting to kill you then simply take a cup of water and add 2 teaspoons of apple cider vinegar in it. Drink it and feel the ease. (if you drink juices instead of water that will affect your fast, regardless, juices can be taken after the fast. They will help you gain your energy).

It doesn't really matter what kind of option you want to choose-out from all these. You just need to stick to whichever you'll choose. Stay determined with your choice and schedule so you can see the changes and effects. It helps you with perfectly up to 80% of transformation.

Here's some information and help regarding personality, if you are unable to decide an option for yourself:

Every type that is mentioned has its own unique advantage. It has its special trait which can be handled according to every user. Hence, if you are a person who sleeps a lot, you can easily go for option number 2. Similarly, if you are always busy in your own world, you can even go for option number 4. If you are someone who loves eating, a foody perhaps? And you don't want to stop try going for option number 3. If you are indefinite in your decisions and you are a person who will never be able to make one decision at a time. Then you might like option number 1. Maybe you are a person who eats a lot at a meal, all to your heart's content? Now there's no room for another meal? Why don't you skip it? You will love option number 6. Are you someone who is a juice lover? And whatever may happen you cannot leave your juice? Then maybe you can try option number 9. Are you like so fat that it is hard even to walk? Then go for option number 7 or option number 8. Try your best! If you think you are none of these and you still don't understand maybe, then you can try option number 5. If after some time you think that you can still not get affected by a particular option? Then sure, you can always try another one. But remember, you need to stick to a single option for a long time (at least a month) to see changes and transformations.

There are some precautions still. People who have medical treatments going on should not consider doing these methods. Pregnant women should not even try to fast. They should eat as much as they want and can. Then we have people who have food disorders and can't stay hungry for a long time due to medical issues, or medical history should avoid intermittent fasting. It is better for these people to exercise and keep eating and stay fit through the gym or yoga. Staying hungry can only create another issue for you if you are one of these people with a medical background. Similarly, if you feel any issue happening to your body after this, then your body can't possibly tolerate it if such a case happens to consult with your doctor. It might be you are weak, or maybe there's some other issue which is not regarding intermittent fasting still, though you can't even know unless you consult it with your doctor.

Chapter 3

Importance of intermittent fasting

When it comes to your body, then you always want to know what the importance will be and how it will affect you. There are many advantages of intermittent fasting, just like the many surveys and research that was conducted on animals in laboratories. It was found that when animals were given foods on their specific time, they stayed fit and healthy. Those animals which were left free to eat as much as they wanted any time started having diseases. This was then converted to a report, and it was finalized that time scheduled eating was a great benefit for health. Hence it was decided to implement on humans.

People who eat without a schedule appeared to have more issues, and most of all, they had obesity in them. People who followed intermittent fasting showed remarkable signs and changes. Through research, it was found that the benefits of intermittent fasting included:

- A proven weight loss without the hassle of going to the gym or exercising.
- A lot of health benefits reducing diseases risks even cancer risks

• Proven brain health and mental stability

1. Reduction of type 2 Diabetes

Some studies show that even type 2 diabetes is reduced from intermittent fasting. It can even benefit the patients that have increased the risk of diabetes by lowering their diabetes level. It provides stability in their blood flow and immune system and helps the digestive system stay healthy. This results in a healthier body.

2. Beneficial to heart

Researches made in laboratories showed that intermittent fasting helps the cardiovascular system. A report in 2016 was conducted and proved that intermittent fasting reduces all kinds of issues like blood pressure, triglycerides, cholesterol, and even heart rate in animals as well as humans.

3. Healthy Brain

Intermittent fasting shows that mental stability is possible, and it provides perfect brain health. Through this, a person stays fresh mentally; he does not feel extra burdened and stressed. Nor does his brain feels tired. A healthy and fresh brain is the achievement of many successes and solutions to many problems.

Your thoughts will be clear. You will stay in peace; you will have the ability to analyze things at a quicker pace. This is best for students as well as for people who have to workloads in offices. It helps them stay off the alert notice and helps them become successful. For students, it might be perfect in learning everything and stay at the top of the class (that is of course, only if you want to).

It gives you the ability to increase the power of your memory too! You can remember things at greater detail. You won't be disappointed by your own brain becoming a vault.

If you are someone who gets depressed a lot? Or, is your teen upset most of the time? Get yourself an intermittent fasting plan and stay healthy and fit. It relaxes your souls and body. It eases your mind. And it is perfect for hasty teens. There won't be any more depression affecting you. You might become just the perfect personality you always wanted to have. By just taking control over yourself, your brain, your soul, and your body.

4. Reduction of cancer

Studies showed that intermittent fasting reduced cancer even in animals. It later proved that humans were facing problems of cancer and the major reason for that was their diet plans. Hence following a planner will reduce the risks of cancer growing in your body.

The main reason for this is that obesity is the major reason for the growth of cancers most of the time. Through intermittent fasting, one loses weight and reduces the chance of obesity. This further reduces the chances of all issues or diseases that might grow in your body.

5. Boosting of biological processes

Biological processes play a massive role in our body's functionality. When you know how to boost your functions, your body interacts with matters at a greater rate. In order to be healthy, you need to know the answers to the questions, whether all hormones or chemicals in your body are balanced or not? If it is not like that then, there might be somethings imbalance in your body, which can cause any kind of illness or cellular degradation, which further leads to premature death! Hence, we need to take care of these questions and find solutions. The best solution is to perform intermittent fasting. This process will keep all your processes stable and in perfect order. It will also boost your biological systems, making you active in every activity that you want to perform or operate.

As it is said, people do not use their brains even up to 1%. The main reason for that is, they are stressed. Their entire system is stressed, not just their mind! This is the cause of eating at irregular times. Thus, if you set your schedule, you'll be able to control yourself rather than being controlled subconsciously.

When a person is under intermittent fasting and if he chooses to not eat for a longer time, then the body of that person goes under the autophagy process. The autophagy process, as mentioned in chapter 2, is a process in which the old cells are cleared, and new cells take place. This process is only possible when a person stays without eating food for a certain period. It helps to ingest the components of the old cell and grow new ones. The new ones can only grow in the absence of old cells. When a person eats something, the glucose can create hindrances in the autophagy process hence try to drink only water or herbal tea during your fast.

Here are some important reasons showing you why intermittent fasting is important for you:

A study went on for 16 weeks, which was conducted on 283 people. Later it was found that people who ate breakfast, and people who didn't eat breakfast had no difference. Basically, it is considered a myth when people say breakfast is very essential, and if you don't eat it, you'll become weak or have diseases. The report concluded that people observed children performing better who had taken their breakfasts. Well, what about those who didn't eat their breakfasts? Some kids perform equally well without any problem. The reason is that it completely depends on the body of what it needs. If it can work even without breakfast or not. It is not essential, though.

There was another survey which denied everyone's assumption that eating more boosts your metabolism rate and burns fats. Well, this process is basically known as the thermic effect of food. It does not depend on how much you eat. It depends on how much calories you take in. For instance, if you are eating six times a day consisting of 500 calories, that's a lot excessive. But if you eat 3 meals consisting of 1000 calories, then it's the same deal. It will burn 300 calories in both ways. Hence it does not depend on the number of meals. It depends on your calorie intake.

People even believe that if they eat frequently, it will reduce their hunger. However, it is another myth, and it is completely wrong. Just like sleeping cycles, the more you sleep, the more your sleep will increase. The more you eat, the more your hunger will increase. It will not reduce; rather, it will make your body used to eat more without any hesitation. So, if you think that eating more will boost your metabolism rate, you are wrong. There's no difference. Studies showed that whether you eat three times or even 6 times a day, your metabolism rate depends on your body. Eating frequently will definitely not boost it.

One other common misbelief among people is, our body needs glucose for our brain to function with improvement. Our brain indeed needs glucose, but, that does not mean you eat carbs as much as you can to boost your brainpower. If your brain needs glucose, then your body will produce glucose itself for your brain. This process is known as gluconeogenesis. Even if you are starving for three days, your body can still produce the required glucose for itself. Hence you don't need to do your best in eating sweet stuff all the time. That doesn't mean not eating either. If you are on a long intermittent fasting process and feel dizzy, it is better to eat something light right away.

Some people even fuss over a myth known as protein distribution over meals. They believe that their body can only handle 30 grams per meal. Well, that is again incorrect. Protein intakes are not depended on a number of meals. It depends on how much protein you eat for your body to handle. And the metabolism rate definitely stays unaffected from protein intakes.

One other common disbelief that is shocking itself is, people, say that intermittent fasting will make you lose your muscle mass. Strictly speaking, these people do diet instead of intermittent fasting, and that is what reduces their muscle mass. Dieting is never healthy. Even bodybuilders are following the methods of intermittent fasting to help them maintain their muscle mass. So, it is clear that it will not eat your muscles away, but dieting surely will! A survey was conducted, and the conclusion which shocked people was that intermittent fasting would actually maintain your muscle mass. Hence this method is very popular among the community.

You might have even heard that intermittent fasting will not affect you only make your health go worse. The truth is it can not only improve body functionality; it has proven to increase a person's as well as animal's lifespans. It is basically the healthiest possible way to live by. Studies showed that this process could also increase your immunity and gene longevity. It has also proven to boost the health of BDNF (brain-derived neurotrophic factor), which is basically a hormone that protects against mental disorders.

Some people even protested that intermittent fasting makes you overeat due to extreme hunger. Well, you might indeed eat a bit more at the end of the day. That is because your body will be deprived of insulin levels and norepinephrine. Also, the levels of (HGH) human growth hormones become low. That is why a person normally eats more. But that doesn't make you fat. You actually lose weight. According to a survey, 3 to 8% and 4 to 7 % of people lost belly fat and weight in 2 to 24 weeks by intermittent fasting. People are more than happy with the results.

Don't just jump to conclusions when it's a matter of your health. It's better to consult some sources and do some research yourself. Your body can handle hunger. It's your mind that gets bothered. You only feel hungry when you know that you haven't eaten. This is relatable for people who are mostly busy with their work. They lose track of time to have meals, and they know that they don't feel hungry at all while they are working. When you try this process, don't forget to occupy your brain. At the start, you will have some difficulties because you might have the habit of eating all the time. Hence, stay safe and stay healthy with a diet plan and treating your time as the most important gift. Do everything at the right time. As you all have heard the phrase "Excess of everything is bad and wrong." The same goes for the case of eating excessively. Always maintain yourself while eating. It was also a fact that people who eat by always keeping a portion of their tummy empty, they were always fit and safe from health problems. Intermittent fasting provides you with all these benefits and keeps you safe from all the health issues. You may never need to see a hospital again even (as some reviews of followers).

Chapter 4

Meals and diet plans for intermittent fasting

There are many ways that people adopt when they are dieting. However, intermittent fasting is not only for making you lose weight; it is to make you healthy as well. Hence when you adopt a perfect diet plan of intermittent fasting (that is scheduled eating), then you would surely be able to be healthy and perfect in shape. When you eat your meal with that specific timing, then you would be able to stay fresh all day.

The process of Intermittent fasting is about eating plans, including your dietary protocols that revolve around periods of fasting (no food or some food) and periods of non-fasting. There are a lot of options of intermittent fasting diets, along with several types of intermittent diet meal plans from which you can choose.

There are many benefits of intermittent fasting, including the process of lowering insulin levels, improving brain health, reducing your inflammation, helping you feel more hopeful and prayerful, and possible weight loss. When considering a complete analysis, Intermittent fasting has shown promising results in broad trials involving animal (mice were the first testing objects) and a few limited human trials.

During these days, you either eat no food at all or 25 % of your regular calorie diet. On your reduced-calorie days, most of the men consume 500 to 600 cal. while women mostly intake 400 to 500 calories. You should consider your own body type first rather than focusing on gender and how much women or men intake calories. You should not jump to conclusions and simply limit the number of calories for yourself.

You can use intermittent fasting meals for fat loss. You don't have to worry about it though, that is because, this meal plan is pleasurable, productive, and simple to understand and carry. The perfect dieting strategy to lose all your aggressive fat is through intermittent fasting. Using this procedure, you can get lean quickly without feeling deprived of calories. And you can continuously eat low calories this way.

This process also has benefitted people from a lot of time consumptions in kitchen cleaning or cooking. After applying intermittent fasting in your life, you will see that you will be a lot more energetic through the entire day and have a keen and sharp focus on every task. Hence, you will be blessed.

Your meals should be well set, and diet plans should be made accordingly after analysis. The recipes and food that you include in your daily life should be a balance of everything. Even if you want to eat pizzas, it is not like you cannot. You certainly can. You only need to know about the specific step (of everything) you take possibly are.

A functional medicinal practitioner stated that most of the health issues are born from chronic inflammation. Also, when we are fighting pathogenic bacteria, we look forward to acute inflammation as it is more of a natural and healthy response. However, long term chronic inflammation at times does not subside if the threat further processes to conditions to cancer. If we want to save ourselves from inflammation, the safest side to consider is through natural remedies. That is often taken care through intermittent fasting because you stay without eating for a certain period of time. It is the best way when we consider the easy and helpful method for everyone. This method is liked by people on a broader scale.

As intermittent fasting asks you to eat in chunks at time intervals, you will be eating less food on those days. This actually eliminates a lot of stress from your routine, also from meal prep tensions. Studies Say that this might also increase your lifespan. (because you give rest to your gut from time to time. It helps you live longer potentially and benefits you in many ways. You don't eat whenever you want to even if its only vegies. You have to eat at particular time spans. Following are some befits mentioned.

- Lowers the risk of cancer

- It enhances hearts health

- It improves autoimmune conditions

- It improves blood sugar

- It encourages weight loss

- It curbs your cravings

- It increases cognitive functions

- It improves lung health

- It helps heal your gut

- You lose your weight quickly

- You get a lot of energy.

- You'll see Your metabolism increase exponentially.

- Your appetite will be drastically reduced.

- This plan will naturally train and help your mind and body to live a healthy lifestyle. You can't expect to live a healthy lifestyle if you don't know how to. You will learn this as you go through the stages.

- This isn't an exercise dependent weight loss procedure. You won't have to exercise to lose weight.

- This plan will work for all ages (unlike a lot of weight loss plans out there) BUT it is not for pregnant and nursing mothers or even children. (You MUST consult your physician before going on any diet).

Before you start this IF plan, it's better you talk with a professional to make sure either if you are ready for it yet. Women should be especially cautious because there are certainly mixed opinions on whether or not certain fasting protocols are healthy for female hormone balance. In addition to this, if you have adrenal fatigue or gut health issues, you should proceed with caution. If you've suffered from a history of disordered eating, you'll probably want to avoid fasting.

Once you start your journey in IF plan, you'll most likely find that you feel fuller for a long time and can keep the meals you eat very simple. There are different ways you can fast. Each of the different plans below is partitioned into beginner, intermediate, and advanced steps along with a typical meal plan for every day. The combination of those nutrients will give you the energy you need to enhance the benefits of your fasting journey. Make sure to take into account of any individual food intolerances and use this as a guide for your health care, and adjust from there.

Chapter 5

Breakfast recipes for intermittent fasting

1. The Easy Italian omelette (melt)

Ingredients:

- three large eggs
- 6 to 8 cherry tomatoes
- 1 tbsp+ 1 tsp extra virgin olive oil
- 1 tbsp of fresh chopped basil
- few slices of fresh mozzarella
- 2 slices of prosciutto di Parma
- salt and pepper, to taste

Instructions:

1. pour 1 tbsp olive oil into an omelet pan, place it over medium heat.

2. While heating, quarter the tomatoes, shred the basil, chop the prosciutto and mozzarella into small pieces.

3. Break eggs into a bowl, season to taste, and whisk until frothy, then pour them into the heated pan. Leave it to cook for a minute, then run a spatula gently around its underside.

4. Cook until top and center look almost set, then scatter prosciutto, mozzarella, tomatoes and basil over one half of the omelet.

5. Fold the omelet over fillings, turn the heat off and leave to sit for a minute. Drizzle remaining 1 tsp of olive oil.

6. Slide omelet on to a plate and eat while hot.

2. Low Carb California Omelette

Ingredients:

- six large whisked eggs
- ¼ tsp sea salt
- ¼ tsp lemon juice
- ¼ tsp hot sauce (you can try Sriracha Sauce)
- two slices cooked bacon
- 3 tbsp butter, ghee or duck fat
- 10-12 pcs cooked shrimp, peeled + deveined
- 2 tbsp minced parsley/cilantro
- ¼ cup diced red bell pepper
- one medium green onion, sliced
- one large avocado, sliced

Instructions:

1. put 1 tbsp olive oil into an omelet pan, place it over medium heat.

2. In a small bowl, whisk the eggs, hot sauce, lemon juice, and salt together.

3. Heat butter in a large nonstick pan over medium-low heat. Once melted pour in whisked eggs. Cook lifting the edges with a spatula and by tilting the pan to allow uncooked egg to run under omelet until set but still moist on top.

7. Arrange the shrimp, bell pepper, green onion, parsley, avocado and bacon across the top of the omelet. Gently fold it in half, cook for another 2-3 mins until cooked through.

8. Serve

3. Low Carb Veggie Bowl

Ingredients:

- few basils leave
- 3 red, yellow/orange baby peppers or 1 small bell pepper
- 1 tsp ghee/extra virgin olive oil
- pinch of sea salt to taste
- 1 tsp flax seeds
- 1 tsp pumpkin seeds
- 1 tsp sunflower seeds
- 1 tbsp butter, ghee/extra virgin olive oil
- ¾ cup shredded kale/spinach
- 1/3 cup sliced shiitake/white mushrooms
- three slices halloumi cheese
- 1 tbsp homemade Low-Carb Marinara Sauce

Instructions:

1. Preheat oven up to 180°C/355°F (fan-assisted) or 200°C/400°F (conventional). Place peppers on your baking tray and drizzle with olive oil and pinch of salt. Roast in the oven for 25 minutes.

2. Place seeds on another baking tray and roast in the oven for 4 minutes until golden. Remove it from oven then allow it to cool.

3. Heat butter on medium heat in a non-stick pan, add mushrooms and cook for 2

minutes. Add kale and cook for a further 2 mins. Season with salt to taste.

4. Fry halloumi in 1 tsp of ghee/olive oil over medium-low heat for about 2 mins per side, or until golden.

5. Once peppers are cooked, allow cooling slightly. Remove stalks and scoop out the seeds.

6. Smash the avocado with a fork and mix with olive oil, salt, pepper, lime, and chili flakes.

7. Place kale and mushrooms in your bowl, along with the seeds, halloumi, peppers, and top with smashed avocado, marinara sauce, and fresh basil.

4. Low Carb Salmon and Cheese

Ingredients:

- 3 tbsp olive oil
- 2 tsp fresh lemon juice
- two medium salmon fillets
- sea salt+ black pepper, to taste
- 2 tsp fresh lemon juice
- 1 tbsp extra olive oil/ghee
- ½ small cauliflower (makes 2 cups cauliflower rice)
- ½ cup shredded red cabbage
- ¼ cup chopped sugar snap peas
- 1/3 cup chopped red pepper
- ¼ small red onion, finely sliced
- 1 tbsp pomegranate seeds (Optional)
- ¼ cup chopped fresh parsley
- 2 tbsp chopped fresh mint
- ½ cup crumbled feta

Yogurt dressing:

- one heaped tbsp Greek yogurt full-fat 5%
- 1 tbsp chopped fresh basil
- 1 tsp fresh lemon juice

Instructions:

1. Preheat oven up to 180°C/350°F (fan-assisted) or 200°C/400°F (conventional). Line a baking tray with greaseproof paper. Season salmon with salt, pepper, and tsp of olive oil. Place on a tray (skin-side up) and roast in oven for 25 minutes until cooked through and skin is crisp. Optional: pan-fry.

2. Blitz the cauliflower florets using the S blade or a grating blade of your food processor until it resembles a rice size consistency.

3. Place it in bowl and microwave on high for 4 mins. Remove from microwave and allow to cool. (Optional: transfer to a muslin cloth and squeeze out the excess water.) This makes it fluffier. Fluff with a fork.

4. Mix olive oil, lemon, salt, and pepper in a small bowl.

5. Add chopped red cabbage; sugar snaps peas, red onion, red pepper, fresh herbs (optional:

pomegranate seeds). Add olive oil dressing and half of the feta to cooled cauliflower rice. Toss to combine.

6. Mix basil yogurt dressing ingredients in a small bowl. Place tabbouleh in your serving bowl.

7. Top with roast salmon, remaining feta, basil yogurt dressing.

5. Low Carb Caprese Omelette

Ingredients:

- three large eggs
- 1 tbsp butter/ghee
- 1/3 cup cherry tomatoes, halved
- 3-6 basil leaves, chopped
- two slices fresh mozzarella
- 1 heaped tbsp grated Parmesan/other Italian hard cheese
- 1 tbsp pesto (homemade pesto allowed)
- sea salt+ pepper, to taste

Instructions:

1. In a bowl whisk eggs together with 1 tbsp water.
2. Heat butter in a small nonstick ceramic skillet at low heat.
3. Pour eggs into the skillet pushing the cooked edges towards center, cook and tilt the pan so that uncooked egg reaches the hot pan.
4. When the top of eggs are set then place half of the tomatoes, parmesan, basil, and mozzarella on one side of eggs.
5. Fold in half then put it on a plate. Drizzle with pesto and top with remaining tomatoes. Optional: drizzle by balsamic vinegar and olive oil. Serve immediately.

6. Sausage and Guac Stacks

Ingredients:

Quick guacamole:

- 1 tbsp butter/ghee
- one medium avocado
- 1/2 small white/yellow onion, chopped
- 2 tbsp fresh lime juice
- salt and pepper to taste

stacks:

- two eggs, large
- 1-2 tbsp ghee
- 170g gluten-free sausage-meat (Italian)

Instructions:

1. Prepare the quick guacamole first. Halve the avocado and scoop it into a bowl. Add the lime juice, salt, pepper, and onion. Mash using a fork and set aside.
2. Heat a pan greased with half of the ghee over medium heat. Use your hands, and create patties from sausage meat. Place it on pan and cook undisturbed for 2-3 minutes. Flip the other side and cook for 1 to 2 mins and set aside.
3. Grease the pan with remaining ghee and crack in eggs. Cook until egg-whites are cooked thorough and egg

yolks are still runny. If you use egg mold, then lower the heat, as it will take longer to cook through.

4. When done, top each patty with your prepared guacamole and a fried egg. Season with salt, pepper to taste and eat immediately.

7. Kale Eggs and Sausage Crumb

Ingredients:

- four eggs, large
- 1 tsp extra virgin olive oil/ghee
- 2-3 good-quality gluten-free sausages
- 1 cup kale, chopped
- 1 tbsp butter, ghee/duck fat
- Salt, pepper to taste
- fresh herbs like parsley
- ½ avocado, medium

Instructions:

1. Remove skin from sausages. Heat 1 tsp olive oil in your pan. Gently sauté the sausage meat on medium heat, breaking it up with a wooden spatula until it's cooked through and resembles a thick crumb. Add the kale and chili or paprika (your liking) and sauté for one further min.

2. Heat a little butter in a pan on medium heat.

3. Crack the eggs into a cup, season with salt, pepper and whisk with a fork. Add eggs to a pan and scramble for 1 min, stirring the eggs with a wooden spoon to prevent sticking. Remove from the heat and let it firm up.

4. Place eggs on a plate and top with the avocado, sausage crumb, and chopped fresh herbs.

8. Fat-Burning Vanilla Smoothie

Ingredients:

- two egg (large) yolks
- ½ tsp pure vanilla extract
- ½ cup mascarpone-cheese (full fat)
- 1 tbsp coconut oil
- ¼ cup of water
- four ice cubes
- 1 tbsp So Nourished powdered erythritol sweetener/3 drops liquid Stevia
- Optional: whipped cream (topping)

Instructions:

1. In your blender, combine water, egg yolks, mascarpone/creamed coconut milk, ice, MCT/coconut oil, vanilla and erythritol, Swerve/stevia. Pulse until smooth.
2. Top your drink with whipped cream and enjoy!

9. Spinach & Feta Omelette

Ingredients:

- three eggs, large
- 1 garlic clove
- 1 cup white mushrooms, sliced
- three fresh cups spinach or 2/3 cup frozen and thawed
- 1/3 cup feta cheese, crumbled
- 2 tbsp ghee
- Salt, pepper to taste

Instructions:

1. prepare the filling. Peel and finely dice garlic and place on a pan greased with tbsp of ghee. Season with salt and cook over medium-high heat for a minute until fragrant. Add sliced mushrooms and cook for 5 mins until lightly browned stirring.

2. Add spinach and cook until wilted for 2 mins (squeeze out the water if using frozen, thawed spinach). Take off and place in a bowl. Discard excess liquids before using the pan for cooking an omelet.

3. Crack those eggs into a bowl and mix with a fork. Season with salt, pepper to taste.

4. Pour the eggs in a hot pan greased with ghee. Use a spatula to bring in the egg from the sides to center for the first 30 seconds. Tilt pan when needed to cover it with eggs. Lower the heat and cook another minute. Don't try to cook it fast or your omelet will end up being too crispy-dry. It should be soft and fluffy.

5. When the top is cooked almost, add the mushroom topping, spinach, and crumbled feta. Fold the omelet in 1/2, cook for another minute, and serve.

10.   Blackberry Cheesecake Smoothie

Ingredients:

- ½ cup blackberries, fresh or frozen
- ¼ cup full-fat cream cheese/creamed coconut milk
- ¼ cup heavy whipping cream or coconut milk
- ½ cup of water
- 1 tbsp MCT oil/extra virgin coconut oil
- ½ tsp sugar-free vanilla extract or ¼ tsp vanilla powder
- Optional: 3-5 drops liquid stevia

Instructions:

1. In your blender, add all the ingredients. Pulse until it's smooth.
2. Pour in a glass and enjoy!

Chapter 6

Lunch recipes for
intermittent fasting

---

1. Mediterranean Low-Carb risotto

Ingredients:

- 1cauliflower (medium head)
- 4chicken breasts, (medium) skinless+ boneless
- ¼cup heavy whipping cream/coconut milk
- ½cup pesto sauce
- 2cloves garlic (mashed)
- zest from ¼ lemon (½ tbsp)
- 2 tbsp freshly chopped – oregano, basil, thyme (1–2 tsp if dried)
- 2tbsp ghee, butter/coconut oil
- pinch fresh black pepper
- ½tsp pink Himalayan/sea salt
- 1cup grated parmesan cheese

Instructions:

1.   Prepare your cauliflower "rice." Remove the leaves and hard center core of cauliflower and cut in florets. Wash

the cauliflower and drain well. Once dry, grate it. Pulse until it looks like rice. A grating blade will make it look real rice.

2. Dice the chicken into medium pieces and place on a pan greased with ghee.

3. Cook for 15-mins. When done, put aside.

4. Peel and mash the garlic. Zest the lemon. Use organic, unwaxed lemons.

5. Grease another pan with remaining ghee add mashed garlic and lemon zest. Cook over medium heat until light golden.

6. Add "cauli-rice," increase the temperature to medium-high and cook for 5 minutes stirring frequently. Time depends on how soft you prefer "cauli-rice." Add the pesto, cream (or coconut milk) and chopped herbs. Cook for 1-2 mins and put aside. Season with black pepper and salt.

7. Add the Parmesan cheese and mix well. Enjoy.

2. Jalapeno Popper Stuff-Burger

Ingredients:

- two eggs, large
- 4Ultimate Keto Buns halved
- 1lb ground beef
- 1cup cheddar cheese
- 4 thin-cut slices bacon, crisped up
- 1/2 cup pickled jalapeños
- 4slices tomatoes
- handful fresh greens of your choice

Instructions:

1. Make the Ultimate Keto Buns

2. Divide the beef into 8 equal pieces. Patty out a piece and fill with jalapeños, cheese, then crispy bacon (you can crisp it in oven/pan).

3. Top with another piece and pinch the sides together sealing the burger.

4. Grill burgers on medium-high heat, each side takes approximately 4 (medium).

5. top with your desired burger toppings. Enjoy!

3. Chorizo-Stuffed Spaghetti Squash

Ingredients:

- 2small/1 medium spaghetti squash (seeds removed)
- pinch salt
- 2 tbsp ghee/lard
- 1small white/brown onion, chopped
- 450g Mexican chorizo
- 1 cup cheddar cheese (shredded)
- 1 cup canned tomatoes (chopped)
- fresh black pepper

Instructions:

1. Preheat your oven first 200°C/400°F. Halve spaghetti squash lengthwise. Use the tip of the chef's knife to break through skin in the middle part and push blade down until it cuts through. Use spoon: scoop out the seeds and discard or reserve for snacking (can roast them). Brush the inside of each half with melted-ghee and season it with salt.
2. Place spaghetti-squash in the oven and bake for 25-40 mins. Check readiness using a fork.
3. grease a large pan with ghee and cook the onion on medium-heat until lightly browned. Then add chorizo and cook it for 3-5 minutes/until browned. Then add canned tomatoes, season them with salt +pepper and combine. Cook for 1-2 minutes.

4. Add the grated cheddar cheese, mix well. Take it off the heat and fill up each of the spaghetti-squash halves with meat mixture.

5. Top it with cheese, place under a broiler for 3-5 mins/until the cheese is melted and crisped.

4. Creamy-Pesto Tuna

Ingredients:

- 1small tin tuna, oily+ drained
- 1 ½ tbsp paleo-mayonnaise
- 1heaped tbsp full-fat Greek yogurt, coconut yogurt/mayo
- 1 tbsp pesto
- 2 tsp lemon juice
- 1/8 tsp sea salt, to taste

Dressing:

- 1 tbsp extra-virgin olive oil
- ½ tbsp apple cider vinegar/lemon juice
- 1/8 tsp black pepper+ salt

Salad:

- 4 leaves iceberg lettuce
- small tomato, sliced
- ½ cucumber, sliced diagonally
- ¼ avocado, thinly sliced

Instructions:

1. Make creamy tuna by mixing all ingredients in a small bowl and mashing together with a fork until combined. Add salt.
2. For the dressing, add ingredients to a small jar and shake to combine.
3. Layer the lettuce, cucumber, tomato in a bowl.
4. Top it with tuna-mix and the avocado. Drizzle it over the dressing.

5. Beef Hot Pockets

Ingredients:

- ½ brown onion, small
- 2garlic cloves
- 1tsp ghee/butter
- 300g ground-beef
- 1-2 small chili peppers, chopped
- 1tsp coconut aminos
- 1tsp Sriracha sauce
- ¼tsp sea salt, to taste and black pepper
- 1 cup spinach, fresh
- 3/4 cup low-moisture mozzarella, shredded
- 1/3 cup almond-flour

Instructions:

1. Preheat your oven first 200°C/400°F (conventional), or 180°C/355°F (fan-assisted). Chop onion and garlic. Heat ghee in a non-stick or cast-iron-pan on medium heat.
2. Add the onion and fry for 2 mins until soft. Add the garlic for 30 seconds. Add beef then cook for 5 minutes until cooked through, breaking the mince with a spatula until fine.

3. Add the chili, sriracha, coconut aminos, and season to taste. Stir through spinach, cook for 1-2 mins until wilted. Turn off heat and put aside.

4. Melt the mozzarella in the microwave for about 60 seconds until it melts. Add almond flour then mix to combine to form a dough.

5. Roll between two sheets of greaseproof-paper/one sheet and silicone mat.

6. Place chili-beef mixture in the center and fold to seal the dough.

7. Careful prick or slice a few air-holes in the top.

8. Place on a greaseproof lined baking-tray and bake in the oven for 15-20 minutes until golden.

6. Pork-Chops with Asparagus and Hollandaise

Ingredients:

- ½ cup blackberries
- ½ cup butter/ghee/extra-virgin olive oil
- 3egg yolks, large
- 1tbsp lemon juice
- 3pork loin chops, bone-in/boneless pork chops
- 2tbsp ghee/lard
- 300g asparagus spears
- Salt, pepper, to taste

Instructions:

7. Prepare one-minute Hollandaise first. Place ½ cup butter/ghee into a wide-mouthed jar, with enough room for a hand blender to fit. Melt butter in microwave.

8. Add the egg yolks and lemon juice. Place the hand-blender in the bottom of the jar. blitz until well combined. season, if required.

9. Heat a frying-pan on med-high heat and melt remaining ghee. Cook the pork-chops for 6 mins on each side and rest for 5 mins.

10. bring 1 pot of water to boil and then blanch the asparagus till 5 minutes. Remove it from water and drain well.

11. Serve pork-chops with asparagus-spears placed over them, and drizzle the hollandaise over top.

7. Caprese-Chicken bowl

Ingredients:

- 1small skinless, boneless chicken-breast
- 1tbsp extra-virgin olive oil
- 1tsp balsamic vinegar
- 1tsp Italian seasoning
- sea salt+ pepper, to taste
- 2 cups spinach/any greens
- ¼ loose cup basil-leaves
- 6 baby mozzarella balls/fresh mozzarella, sliced
- ½ avocado, thin sliced
- 1/3 cup cherry-tomatoes, halved

Instructions:

1. Place the chicken breast in a sealable container with olive oil, vinegar, Italian-seasoning, salt. Put it in refrigerator to marinate (10 minutes to overnight).

2. Heat pan over medium-high heat. Transfer your chicken along with the marinade to the skillet and sear 4-5 mins per-side until chicken is cooked. Transfer to a cutting board and slice.

3. In a jar, whisk the dressing ingredients. Set aside until ready to serve.

4. Assemble the salad by putting the veggies and mozzarella in a bowl.

5. Top with chicken and dressing. Serve.

8. Portobello-Mushroom Mini Pizza

Ingredients:

- 1skinless(small), boneless chicken-breast
- 2large portobello-caps stems removed
- ½ cup pesto
- 10 black-kalamata olives
- 1tbsp canned peppers
- 1tbsp capers
- 1 cup shredded Italia-blend cheese

Instructions:

1. Preheat oven 190°C/375°F (conventional), or 170°C/340°F (fan-assisted) and place the mushrooms on a baking sheet. Divide the pesto between the mushrooms.
2. Fill centers with cheese then top with your desired toppings.
3. Bake till 10-15 minutes, till the cheese is bubbly and mushrooms are starting to soften.
4. Serve. optional: sprinkle with red pepper flakes

9.   Salmon Patties

Ingredients:

- 500g tinned salmon, drained
- 2eggs, medium
- 50g almond flour
- 2tbsp chopped parsley
- salt/pepper, taste
- 1tbsp chopped dill
- 1tbsp lemon-juice
- 1onion (small) diced
- 1clove crushed garlic
- 1tsp paprika
- ½tsp cumin
- ½tsp turmeric
- 2tbsp ghee/lard/duck-fat for frying
- 1small-avocado
- ¼cup mayonnaise
- 1tbsp lime-juice
- 1clove garlic-crushed
- 1tbsp parsley/cilantro
- salt/pepper to taste

Instructions:

1. Place the salmon patty ingredients (except ghee/lard/duck fat) in large mixing-bowl and combine.

2. Scoop some of the salmon-mixture into a ¼cup-measuring. With the back of the spoon, press the mixture into the cup to pack it. Turn measuring-cup over onto plate. Repeat to make 12 patties.

3. Heat oil in frying pan and cook the salmon patties on a gentle heat. Cook for 4-5 mins (each side) and use spatula to flip.

4. While patties are cooking, prepare the avocado-dip. Place all ingredients in a blender. Process until smooth. Serve.

10.   Almond-Maca Smoothie

Ingredients:

- 3/4cup unsweetened almond milk
- ¼cup coconut-milk
- 1tbsp unsweetened almond butter
- 1tbsp MCT-oil/Brain Octane Oil/extra-virgin coconut oil
- 1tbsp collagen-powder
- 2tsp maca-powder

Instructions:

1. Mix all the ingredients in a blender
2. enjoy!

Chapter 7

Dinner recipes for intermittent fasting

1.   Salmon-Asparagus &Blender Hollandaise

Ingredients:

- 1tbsp avocado oil/ghee/extra-virgin olive oil
- 2small wild-salmon fillets
- small bunch asparagus
- two egg yolks
- 6tbsp unsalted butter/ghee, melted
- 1tbsp lemon juice
- Salt &pepper
- pinch of garlic powder, cayenne pepper, onion powder/paprika
- dash of water if too thick

Instructions:

1.   Heat oil over medium-high heat in cast iron skillet first. Season salmon with salt pepper and place skin-side down in the skillet. Sear for 4-5 mins till salmon easily releases from the bottom of pan.

2.   Flip continue to sear another 4-5 minutes until it releases.

3.   Flip. Place the asparagus in a skillet, cook for 3-4 mins tossing a few times. Set aside.

4.   For the hollandaise sauce, heat butter over medium heat until melted and bubbling. Then take it off.

5.   Place egg yolks in a blender with lemon juice and cayenne. Blend 30 seconds until yolks are broken down. The sauce should thicken till then. If it's thick, add a dash of water then blend again.

6.   Season with salt, pepper, cayenne pepper. Pour over salmon, asparagus and serve.

2. Butter Braised Cabbage with Crispy Ham

Ingredients:

- ½head white/green cabbage
- 2sticks unsalted butter
- sea salt +black pepper (to taste)
- 6slices prosciutto di Parma

Instructions:

1. Slice cabbage. Place it in a dutch-oven/large saucepan.
2. Chop butter in chunks and place on top of the cabbage.
3. Put the lid on pot and cook on low gas for 2hours, stir 15-20 mins. Do not put water in.
4. Heat oven for 180°C/355°F (fan-assisted), or 200°C/400°F (conventional). Place prosciutto on an oven tray and cook 10-15 mins until crispy.
5. Cool, and crumble roughly into a container.
6. Once the cabbage is finished, serve with topping of black pepper and crumbled prosciutto.

3. Salmon & Avocado Omelette-Wrap

Ingredients:

- three large eggs
- ½average avocado
- ½package-smoked salmon
- 2tbsp full-fat-cream cheese
- 2tbsp freshly chopped chives
- 1medium-spring onion
- 1tbsp ghee/butter
- sea salt +pepper to taste

Instructions:

1. Crack the eggs into a bowl. Add a pinch of salt; pepper then beat them well with a whisk/fork.

2. Mix cream cheese along with chopped chives. Slice smoked salmon, then peel and slice the avocado.

3. Pour eggs evenly in a hot pan greased with ghee. Cook on medium-low heat. Don't rush it. Use a spatula to bring the egg from sides to the center. Cook another minute or two.
4. Slide the omelet onto a plate and spoon the cheese spread all over it.
5. Add the avocado, salmon, chopped spring-onion and fold-into a wrap.

4. Bun-less BBQ Guac Burger

Ingredients:

- 500g minced beef
- 1tbsp Dijon mustard
- 1tbsp freshly chopped thyme/1tsp dried thyme
- 1tbsp freshly chopped oregano/1tsp dried oregano
- 2cloves garlic, crushed
- ½teaspoon salt to taste
- Freshly-ground black pepper
- 2medium-heads iceberg lettuce

Instructions:

1. Start by Guacamole. When done, cover it with foil to prevent the avocado from browning. Place the minced beef in a bowl. Add thyme, mustard, oregano, garlic, salt, and pepper.

2. Combine them well using your hands. Divide the mixture in 4parts and form 4large burgers.

3. Place your burgers on the barbecue and cook on high flame for 8-10 minutes. Flip the burgers halfway through cooking-time.

4. Fold the iceberg-lettuce leaves in serving bowls, add guacamole and top it with burgers. Serve.

5. Crispy Lemon &Thyme Chicken

Ingredients:

- 8chicken (boneless) thighs
- 1tbsp freshly chopped thyme/1tsp dried thyme
- 2tbsp fresh lemon-juice
- 1tsp fresh lemon-zest
- 2cloves garlic, minced
- 2tbsp extra-virgin olive oil
- 2bsp ghee/lard/coconut oil
- 1tsp salt
- ¼tsp freshly-ground black pepper

Instructions:

1. Use a sharped knife or kitchen shears and cut out bone.

2. Place the thighs on chopping board (skin side up) and use a meat pounder to flatten it so thickest

parts can cook. Layer the thighs in a bowl then add the seasoning (lemon juice, lemon zest, olive oil, thyme, minced garlic, salt, pepper). Mix to cover evenly. Put in fridge for least an hour/overnight to marinate.

3. Remove from fridge and place thighs on a paper towel to remove moisture. It's better if you will remove all the spices from the skin apart from meaty side to prevent it from burning.

4. Heat a large skillet. Grease it with ghee on medium-high heat. Place the chicken. Cook undisturbed for 7-10 minutes. Rotate the pan halfway to ensure even-cooking. Turn the chicken thighs to another side. Cook for 2-3 minutes until cooked through.

5. Put it on a baking sheet so juices can drip down. Let thighs rest for a few minutes before serving.

6.   BBQ Meatball Skewers

Ingredients:

- 600g ground-beef
- ½medium red onion
- 2cloves garlic, crushed
- one egg
- 1tbsp paprika
- 1tbsp fresh oregano/1tsp dried oregano
- 2tbsp fresh basil/1tsp dried basil
- 1tsp fresh lemon-zest
- salt to taste
- 200g Spanish chorizo

Instructions:

1. Peel and dice the onion. Put meat in a bowl first and add crushed garlic, onion, lemon zest, chopped herbs, paprika, and egg.

2. Combine all the ingredients. When the meat is ready for a barbecue, slice the chorizo sausage. Using your hands, form 24 meatballs:

3. Assemble skewers by piercing through the meatballs and chorizo slices in alternating order (3 meatballs and four chorizo-slice per skewer).

4. Place skewers on the barbecue and cook for 7-8 minutes/until crispy. Enjoy!

7.   Italian Meatza

Ingredients:

- 1.1pound ground-beef
- 1tsp dried oregano and basil
- ½tsp salt +pepper
- Toppings
- 2cups wild mushrooms
- 2tbsp ghee/butter
- 2cloves crushed garlic
- 1package spinach
- 2tbsp pesto sauce
- ¾cup shredded-mozzarella cheese

Instructions:

1. Place chicken breast in a container with olive oil, salt, vinegar, Italian-seasoning. Put in refrigerator (10 minutes to overnight).
2. Preheat your oven to 400°F.
3. In a bowl add oregano, ground beef, basil, salt, pepper, and mix.
4. Using your hands to form pizza "crust" (half an inch thick).
5. Put it on a baking sheet with parchment paper.
6. Place in the oven for 10 mins.
7. prepare the toppings.
8. Slice your mushrooms.

9. Heat a pan greased with the ghee/butter. add the crushed garlic. Cook for 1minute.

10. Add mushrooms then cook for 5 minutes, stirring.

11. Toss in the spinach and cook for one more minute.

12. Season with salt &pepper.

13. Remove pan from the heat.

14. When the meat crust is cooked, remove it from the oven and spread your pesto sauce on top. Add ½ mozzarella cheese, spinach, and mushrooms.

15. Finish remaining cheese and return to oven for 5 minutes/until cheese is melted.

8. Salmon with Creamy Spinach Hollandaise

Ingredients:

- 1small salmon/trout fillet
- ½ large packet spinach
- 1tbsp heavy-whipping cream/coconut milk
- 2tbsp ghee/coconut oil/extra-virgin olive oil
- 1serving Hollandaise sauce
- salt &black pepper to taste

Instructions:

1. Preheat your oven 200°C/400°F. Put salmon in a baking tray and drizzle with half olive oil/ghee/coconut oil. Season with salt, pepper and put in the oven. Cook till 20-25 minutes.

2. Wash spinach and place in a salad spinner to remove excess water/pat dry with a paper towel.

3. Grease a skillet on medium-high heat. Add spinach and cook for 3-5 minutes, mixing. Season it with salt.

4. Add heavy whipping cream.

5. Take it off the heat and set aside. Prepare the Hollandaise sauce.

6. Remove salmon from the oven and put aside for 5 minutes.

7. Place creamed spinach on a serving plate and top it with baked salmon.

8. Pour on the Hollandaise sauce. Enjoy!

9. Lamb Souvlaki

Ingredients:

- 1.8lb/0.8kg lamb leg/shoulder (boneless)
- handful chopped mint
- 2tbsp fresh-chopped rosemary
- one lemon, juiced
- ½cup extra-virgin olive oil
- 1/2tsp sea salt/pink Himalayan salt to taste

Instructions:

1. Dice the meat to medium-sized pcs.
2. Put in a bowl and add oil and juiced lemon. Chop the mint and rosemary and add to bowl. Season with salt. Mix well. Put in the fridge for 4-8 hours or overnight. Mix to avoid drying.
3. Preheat the oven to 230 °C/ 450 °F. Pierce the skewers through each meat cube and place them on a rack and in the oven.

4.   After 10-15 minutes, turn each of the skewers to other side and cook for 5-10 minutes.

5.   When crispy browned, remove them from oven and let it cool.

6.   Serve.

10. Chocolate-Coconut Smoothie

Ingredients:

- ½large avocado
- one ¼cup almond-milk
- ¼cup coconut cream/heavy whipping cream
- 1tbsp flax meal/chia seeds
- one ½tbsp cacao powder
- 1tsp virgin coconut oil/MCT oil
- 1heaped tbsp almond butter/another nut/seed butter

Instructions:

1. Mix all the ingredients in a blender. Then add toppings.
2. enjoy!

Chapter 8

Exercises for weight loss while fasting

Can you exercise while fasting? This is a question which everyone hears, and everyone asks. People think that they get energy from the food they eat and therefore, fasting and exercising at the same time will be difficult. Some people have very tough jobs where they have to do lots of physical work. It is difficult for them to fulfill their work demands while fasting. Everybody wants to know the truth of exercise while fasting. Let's consider this issue logically for a while.

When you eat your meal, the insulin level in your blood goes high, and your brain asks your body to use some of the energy instantly. The leftover energy gets deposited in the liver in the form of glycogen or sugar. The liver starts producing fat or DeNovo Lipogenesis when the glycogen stores are full. Dietary protein which we take as food breaks down into its more straightforward form, amino acids. Our body uses some of these amino acids to repair our muscles and proteins, and the excessive amount of these amino acids changes into glucose. Our intestines directly absorb fat it gets stored in our body as fat without any further breaking down.

Intermittent fasting is getting more and more popular day by day because people who have experienced it claim to have satisfactory results. They share that intermittent fasting improved their energy level and helped to maintain it for more extended periods. Their blood sugar levels also became stable without any fluctuation, whereas weight loss was the prime benefit of the intermittent fasting.

With our routine of having snacks or small meals every two hours, it is very logical that our bodies feel refreshed and at home with the patterns of intermittent fasting which our ancestors used to have naturally.

But the question is, what would be the form of exercise while fasting? The answer to this question depends on several factors. The type or duration of fast you choose and how your body responds to it are the two key factors to consider while planning exercising during fasting. People who want sixteen hours fast will have different exercise capacity and requirements than those who opt for alternate day intermittent fasting.

Is it Ok to Exercise While fasting?

If you are fasting for weight loss and your body has enough fat stores, you may opt for exercising while fasting. But you must consider some pros and cons of using and fasting at the same time.

According to research studies, exercising with empty stomach impacts your metabolism and muscle biochemistry mechanism. This process further depends on the stable level of blood sugar and a person's insulin sensitivity. The studies are also in favor of exercising right after eating your meal before the absorption or digestion starts. This factor is even more critical for people suffering from metabolic syndrome or type 2 diabetes.

According to fitness experts and nutritionists, a substantial positive effect of exercising while intermittent fasting is that the stored carbohydrates in your body break down more quickly. As a result, your body will burn more fat from its already existing stores to maintain the energy level for exercise.

If you want to opt for some cardio exercises while fasting to burn more food fat, think about it because there is a downside of it as well. In some experts' opinion, your body may start utilizing your muscle mass protein to produce energy when you exercise in the fasted state. Moreover, in the absence of your regular meal routine, your body will have less energy for exercising correctly.

Some other experts say that the long-term effects of exercising along with intermittent fasting are not very desirable. The process of depleting energy and calories in your body will ultimately slow down your metabolism in the long run.

Making your exercise more effective while fasting

If you are going to carry on with your exercise routine along with intermittent fasting, there are some points to consider which can make your workout more effective.

1. Select the time wisely

In the opinion of expert dieticians and nutritionists, you must consider if you want to exercise before, during, or after your eating time. This decision is essential to make your exercise during fasting more effective.

If you select the LeanGains 16:8 method of intermittent fasting, you will fast for sixteen hours and eat all of your meals within eight hours of your eating time.

But you must choose the timing of exercise according to your physical needs if you feel comfortable to exercise while fasting, go for it. But for some people, it is difficult to perform well in exercise on an empty stomach. They must adjust their exercise routine within their eating window. This way, they would be able to have some post-workout nutrition as well, which is necessary if you are under-weight or trying to maintain weight. But if you are interested in losing weight through intermittent fasting and keep your energy level even after exercising, exercising on an empty stomach is your cup of tea.

2. **Think about your macros to select your workout routine**

The type of food or significant parts of the meals you take before and after exercising, also play an essential role when it comes to select your model and timing of exercise. If you do strength training, you will need more carbohydrates. But you will need more proteins and lesser carbohydrates if you opt for high-intensity interval training. Similarly, you will require a protein-rich diet for weight training as well. So, consider the type of food you also consume to optimize the results of your exercise.

3. Choose appropriate post-workout meals to build or maintain muscles

Some experts are of the opinions that if you want to exercise during intermittent fasting, schedule it during your eating periods. So that your body may repair itself for the workout break down. If you are doing intensive workouts, make sure you eat enough proteins to renovate your body after exercise. The best way to avoid any muscular fatigue or damage after the workout is to have around 20 grams of protein within thirty minutes of workout. Mostly, protein shakes are used for this purpose. But people who opt for swimming or weightlifting, usually have eggs or fish as post-exercise diet.

How can you do your exercise while fasting safely?

You can continue only those exercises for a long time which are safe for your body. You cannot risk damaging your body while exercising. If some movements or postures are painful for your body or you feel over-exhausted after them, quit instantly. Never push the limits of the body too far to avoid injuries during exercise. Follow these tips to stay safe when you exercise during intermittent fasting.

- **Eat before your moderate- to high-intensity workout**

You need to be extremely careful about scheduling your exercise and meal times. Nutritionists and fitness experts say that your body needs some glycogen stores to utilize as fuel during exercise. So, it is advisable to have a meal shortly before exercising.

- **Stay hydrated**

You must keep yourself hydrated while fasting. Many experts do not exclude water from fasting sessions. Staying hydrated is more important for people who have to do heavy physical jobs in harsh weather conditions. Similarly, when you exercise, you sweat a lot. As a result, your body needs more water. Moreover, you will need a water intake to calm down your body after exercising. Dehydration can cause anxiety, low blood pressure, dizziness, and even severe health conditions.

- **Maintain your electrolytes**

Your body needs electrolysis for the proper working of its nervous system and many other functions. This need gets enhanced when your body is not taking in any food for long hours. Coconut water is a good, healthy, and low-calorie drink to replenish electrolytes.

- **Adjust the duration and intensity wisely**

Start with lighter exercises for short periods. Gradually increase the duration and severity of your training as you develop stamina. Never over-do your body because too intense exercise can damage your muscles permanently. Listen to your body. Always start your exercise session by warm-up exercises before starting high-intensity workouts.

- **Select the type of fast appropriately**

Your fasting method must support the kind of exercise you intend to choose. If you decide twenty-four hours pattern of intermittent fasting, low-intensity exercises are suitable for you. You may pick restorative yoga, Walking, or general Pilates. But if you want to opt for 16: 8 method of intermittent fasting, you may stick to a specific type of exercise. Because most of your fasting time goes to sleep and you have enough energy to do whatever kind of activity you want.

- **Make your body the priority**

The most important thing you must keep in mind is to listen to your body while exercising during intermittent fasting. You are dehydrated, or your blood sugar level is low if you start feeling dizzy or week during or after exercising. If this happens with you, quit exercise at once and take a carbohydrate-electrolyte energy drink instantly. Have a nutrition-rich meal as soon as possible. The best way to get the maximum output from your exercise while fasting is to keep yourself hydrated enough to go through exercise sessions without over-exhausting your body. This precaution will prevent such situations and any risk of damage to your health.

Exercising during intermittent fasting may work for some people, but this experience may not be very pleasant or fruitful for others. You must consult your physician or healthcare provider before opting for any diet or exercise. This consultancy becomes even more crucial if you have some chronic health condition like diabetes or cardiac issues. If you are a woman and expecting a baby or a breastfeeding mother, devise your intermittent fasting and exercise plan which does not harm you and your baby's health.

Is it difficult to exercise while intermittent fasting?

During your first days of fasting, you always feel starving in the morning even if you don't exercise. This phase may last for around ten to twelve days, depending on your body, stamina, and eating habits. You may start some engaging exercises during this period to distract your mind from hunger pangs.

Always choose exercises which you find interesting. You are already suffering from anxiety due to fasting, so, never push your limits by choosing activities you find boring. You may play music while exercising to boost your stamina and keep your mind off your empty stomach. Location of your exercise can also help a lot. Go to a park or a ground with a beautiful view. Open-air activities will have a positive impact on your mood, and you will feel less cranky. If possible, exercise with a group or a friend. This addition will boost up your morale, and you will enjoy your exercise sessions.

- **Cardio Exercises While Intermittent Fasting**

When we say that exercising while intermittent fasting affects the hormones positively, it happens because of the destruction of glycogen stores. Although doing cardio while intermittent fasting is alright, yet your routine will depend on how fast your body can burn fat instead of glucose. This quality is called fat-adaption. If you have just started exercising and fasting, your stamina may not be very high. For some people, it may take around six months to get used to this regime. If you are a professional athlete and you don't want to risk your race results, never start a fasted raining a couple of weeks before your competition. Your body may react any negative way to the new routine, and that will cost you fitness. Avoid too long fast before exercising and re-fuel your body aright after finishing the workout. Don't forget to hydrate yourself properly before and after cardio to replenish your body.

- **Intermittent Fasting and High-Intensity Interval Workout**

High-Intensity Interval Workout or Burst training means a minimum of four minutes of intense exercise followed by rest for a total period of only around 20 minutes. This HIIT exercise is not only time-efficient but according to research, it impacts positive impacts on your health in many ways. The benefits which HIIT provides you with are not achievable through jogging or aerobics. The most incredible effect of this type of workout is the increased production of the human growth hormone.

Moreover, HIIT exercises improve your body's composition, activate brain function, reverse the biological clock in your brain and muscles, increase testosterone level, and prevent anxiety and depression. When you do HIIT training during fasting, all these benefits get increased. This method is the optimal way to include in your fasted period. To increase the profits of HIIT while fasting, prolong your fast after two to three hours after the workout.

Is weightlifting advisable While Fasting?

According to experts' opinion, it is alright to do weight training during fasting. But you must give detailed thought to the role of glucose after your weightlifting session. This step is crucial, especially when you are lifting weights in the fasted state.

Your glycogen stores are already used when you exercise during fasting. If you include weightlifting as your daily exercise routine, you can do so on an empty stomach. But to avoid any health risk due to low blood sugar level, plan a meal immediately after your workout. Unlike a HIIT session, weightlifting makes your body demand a protein-rich heavy meal right after the exercise session. During the process of your body is adapted to be a fat burner, weightlifting may cause some negative or unwanted impacts on your body. These effects are usually similar to those which you may experience while doing cardio in a fasted state. In this case, give your body some time to adjust to your new eating routine. You may like to plan your weightlifting sessions right after eating, and for this purpose, you may start your fasting session two to three hours after the workout. Schedule fasted exercises when you do HIIT training.

Best options for exercise during Intermittent Fasting:

- **Light jogging or Walking:** a walk of one to four miles is a good option. Put your headphones on and go on a walk anywhere you like.
- **Dance:** any dance is an exciting type of exercise. It can make you forget everything about hunger. You may like to activate your body through steps of hip-hop, Zumba, or ballet along with music tunes.
- **Yoga:** mostly people choose to practice yoga before eating anything, on an empty stomach. Yoga gives you a light and clean feeling, which lets you concentrate on your body's movement and breathing.
- **Pilates:** just like yoga, people who do Pilates often like to do it on an empty stomach. So, you can easily manage this exercise while fasting.
- **Tennis:** you can very easily manage a morning match of tennis while fasting. It is an excellent overall exercise for your body.

- On the treadmill: try running on an incline of 3.5 mph at an eight incline or higher. The treadmill is a scientifically proven way to lose fat than running or jogging regularly.
- **Cycling:** cycling is a great cardio workout. You may enjoy stationary cycling within your home premises or outdoor cycling through a park. Both the options are equally beneficial.

Workouts to must not do in a fasted state:

- CrossFit
- Boxing
- HIIT – classes such as Barry's or Orangetheory, etc.
- Powerlifting

Plan for Working out While Fasting

On Monday

Go for Intermittent fast from 7 pm to 11 am and incorporate a 30-minute burst training session on an empty stomach at 7:30 am.

Tuesday

Have Intermittent fast from 7 pm to 11 am on Tuesday and as exercise, choose 30-minute High-intensity interval training at 7:30 am while fasted.

Wednesday

Again pick intermittent fasting from 7 pm to 11 am and have a weightlifting session of 60 minutes at noon after eating. But don't go for a heavy post-workout meal. Instead, give it a two- or three-hour's break.

Thursday

Maintain the routine of Intermittent fasting from 7 pm to 11 am but give your body a break from exercise. Light, occasional Walking for a short period can be included if you don't want to be too lazy or disturb your routine.

Friday

Again, fast from 7 pm to 11 am and sweat up with an energetic session of burst training for thirty minutes while fasted at 7:30 am.

Saturday

Along with your routine fast from 7 pm to 11 am, have breakfast at 11 am. Do some low to medium intensity exercise for a longer duration. Hiking or jogging may be a good idea, but you may come up with any suitable option of your choice.

Sunday

On Sunday, change your fasting routine a little and give your body some relaxation. Fast from 7 pm till noon. Have a thirty minutes long session of HIIT on an empty stomach at 9 am.

If you can manage 30 minutes

Use half of this time to walk outside in a park, ground, or along a hiking track through the mountains. You may substitute this outdoor walk with a 1-mile walk on a treadmill or an incline. Do some bodyweight exercises, squats, push-ups, lunges, or core work in the remaining fifteen minutes.

If you can manage 45 minutes:

Walk for twenty to thirty minutes or jog for two miles outside. You can choose spending this time on treadmill or incline if you are not an outdoor person or due to weather conditions.

Do some yoga or weight training for fifteen minutes to give a change to your body.

If you can take out 1 hour:

Have a check if bike ride or swimming classes are available in your area. Going for a long walk in a park can be an option. It will be more fun if you go on a walk or bike ride with a friend. You may go for swimming in the pool or some natural water body if available.

You may enjoy your favorite t.v. Shows or listen to your favorite music while spending this hour on treadmill or incline also. It will be a perfectly productive way of spending some quality time with yourself.

Keep one thing in your mind, the primary purpose of every exercise id to keep your body moving. Choose whatever option suits you and your physical requirements the best. You may opt for yoga, play tennis, go swimming, spend time on incline or treadmill or enjoy a hiking trip with friends. Keep yourself hydrated and quit all the intense physical activities if you feel dizzy or uncomfortable.

Summary

- Exercise in the fasted state is not only alright, but it is precious for hormone optimization. This hormone optimization further leads to many physical benefits.

- `You can combine burst training and intermittent fasting to maximize the benefits of both. This multi-therapeutic approach will lead you to many health benefits, which only fasting or only exercise cannot provide you.
- Although it takes time for your body to get adapted to, yet you can do cardio and weight training in a fasted state.
- Choose to exercise early in the morning while fasting because it is the best way to match the body's natural circadian rhythm.
- You can reap the benefits of fasting even after your workout also unless your workout consists of endurance cardio or heavyweight session.

Chapter 9

Combing Keto Diet with Intermittent Fasting

Intermittent fasting and the ketogenic diet are two diverse dietetic strategies which have many standard features. Both make the body burn fat, and both help you boost ketones. If you combine both of them, definitely you will be in a position to reap the benefits of both of these strategies.

You may witness people who are on a keto diet, incorporating fasting as their dietary strategy as well. Utilizing both of these methods at the same time increases energy level, stamina, helps you tackle plateaus, promotes weight loss, and enhances longevity. These benefits are not random individual experiences, but these are proven scientific facts after research.

You don't' need fasting when you are on a keto diet. If it does not suit you, don't push your limits to fast. Restricting your body's needs non-practically is always a wrong decision, which may cost you your health and well-being. If your body does not accept it and you feel bound, there is no purpose to follow a dietary routine. You must understand the two basic terms called "feeding" and "fasting." Your body is on fast when you don't eat for an extended period, intentionally and deliberately. You are in feeding state if you are taking your meals regularly without skipping your snacks or any of your primary meals.

In this chapter, we will explain how the keto diet and intermittent fasting work in collaboration to improve health and boost weight loss. We will also have a look at ways to integrate intermittent fasting into your keto diet. But first, let's understand what intermittent fasting and keto are and how do they affect our bodies.

Why do people like to combine keto with Intermittent Fasting for Weight loss?

Intermittent fasting and keto both help to weigh loss. But both of these diets involve skipping meals or some particular nutrients so; everybody cannot attempt them.

Do keto and intermittent fasting work better together?

For some nutrition experts, combining keto with intermittent fasting is not a good idea. The level of ketones increases in the body while we fast, and the same happens if we are on the keto diet. While being in a state of nutritional ketosis, our brain will depend less on glucose for energy. That is why the change into a fasted or ketogenic state during the day gradually becomes smooth when you have eaten ketogenic or low carb diet for a few days.

Combining intermittent fasting with the keto diet can maximize the benefits of both according to the expert dieticians and nutritionists. In the next phase, people may cope with their weight loss plateau because they would be consuming lesser calories while fasting. This progression can be natural for those people who feel satisfied after eating vast amounts of fat as part of their meals. A shrinking eating window does not affect such people because they take a significant amount of calories in the form of fat.

Which people can choose an Intermittent Fasting Keto Approach?

People who have adjusted their bodies to the keto diet after being on this nutritional approach for around two weeks and with no severe or chronic health issues can integrate intermittent fasting with their keto diet. When we talk Talking about chronic health issues, the keto diet is favorite among those with diabetes or pre-diabetes. According to healthcare experts, not eating for long periods can be dangerous for some people. It can be especially alarming if you are a patient of eating disorders, chronic kidney issues, or continuing cancer treatment. It may be equally dangerous for breastfeeding or pregnant mothers. In all the above-mentioned conditions, never combine keto with fasting because it will over-shrink your food intake. Even fasting or keto individually are not suggested for these people without having detailed discussions with their healthcare providers. If you achieve your weight loss target with the only keto diet and your body feels good, there is no need to combine intermittent fasting with it.

The Correct Method to Start an Intermittent Fasting Keto Combination Diet

Never start keto and intermittent fasting simultaneously. Your body goes through a massive change while switching to ketones instead of glucose as fuel. It will get even more intense if you start also fasting at the same time. The experts recommend to let your body adjust to the change of diet after starting keto. You may incorporate intermittent fasting when you think you have the stamina to carry on with your routine activities with a limited amount of food intake.

Choose the time for fasting wisely according to the needs of your body. A 16:8 fasting method which consists of sixteen hours fasting and eight hours eating window is considered the best. Twelve-hour fasting is also an option, but usually, everyone spends the night time without eating, so it is not very useful. But as a beginner, you may opt for this option to build your stamina.
In the beginning, try to delay your breakfast and keep extending the time until lunchtime. Your body will get used to skipping breakfast very quickly in this way. When your body accepts this change, restart having your breakfast, and extend your nighttime fasting because breakfast is an important meal, and several people don't feel very active when they skip breakfast. Breakfast is essential for maintaining blood sugar and insulin level, dynamic cognition, and improved metabolism. Health experts suggest to

continue intermittent fasting with a keto diet only for six months and then switching back to a low-carb diet.

A Model Menu for the combination of Keto and Intermittent Fasting

After getting approval from your healthcare providers, when you are ready to integrate keto with intermittent fasting, the most important question is, what to eat and when. Your concern is genuine because fasting is all about timing, and keto is all about the type of food.

Choose any method of intermittent fasting according to your physical requirements. Always discuss your diet chart with a nutritionist or health expert who has a thorough understanding of keto and intermittent fasting.

Day 1

On the first day, have a breakfast of scrambled egg topped with avocado pieces with black coffee at 10 am take water through the entire morning.

Have lunch at 1 pm and have a substantial green leafy salad with a dressing of two tablespoons olive oil, three ounces grilled or baked salmon and vinegar.

As a snack, have one-fourth cup of Macadamia nuts. But if you don't feel like having meals, you may skip it.

Have your dinner at 6 pm you may take one chicken leg with skin, one-fourth cup of wild rice, and two cups of zucchini cooked in olive oil.

Day 2

On day 2, have breakfast at 10 am again. Enjoy pure hot tea, a smoothie of any ingredients which are keto-friendly and of course, a proper intake of water throughout the morning.

You can have three ounces grilled chicken breast in your lunch at 1 pm combine it with cauliflower and broccoli with some olive oil and avocado.

Have unsweetened coconut chips as a snack at 3 pm if you like.

Three ounces of baked tuna is an ideal dinner; you may cook it in olive oil and have it on a bed of Asian coleslaw with a topping of sesame seeds and olive oil.

Day 3

your breakfast on the third day will consist of keto chia pudding and black coffee with a generous amount of water.

You can have omelet of three eggs loaded with one cup of spinach, peppers, and cooked in olive oil. Top it with half avocado and a half cup of sliced tomatoes.

At 3 pm, Olives are an excellent choice for an optional snack.

Three cups of kale salad with three ounces of shrimp drizzled with two tablespoons of vinegar and olive oil are your dinner for the third day.

The Potential Health Benefits of a Keto Intermittent Fasting Diet

One thing is proved that the ketone level in our body increases when we combine intermittent fasting with keto. But there are no researches particularly on combining keto with intermittent fasting. This combination may promote weight loss, but everyone's body responds differently.

Most of the doctor who suggests intermittent fasting and keto to their patients find it helpful for maintaining blood sugar level because fasting lowers the insulin level. This combination makes the burn cleaner fuel i.e., ketone for energy. This change decreases the production of toxic materials in our body and helps Alzheimer patients.

Some doctors themselves practice intermittent fasting to stay fit and reduce belly fat. This reduction further decreases the waist also. Reducing fat from around your belly fat prevents you from heart diseases as well.

Is combining Keto and Intermittent Fasting risky for your health?

Keto diet was developed for the people with a seizure disorder, and they find it a quite pleasant experience with excellent health.

But like every other diet, you must keep an eye on what you are eating. Poorly planned keto with lack of important nutrients can cause malnutrition and overall poor health. Taking an appropriate amount of food consisting of quality ingredients becomes necessary when you are fasting along with the keto diet. Fasting already limits your calorie intake to a specific time, and if you don't take adequate amount of food within your eating window, your body may suffer. You may face drastic weight loss, which is unhealthy or loses your muscle mass. This condition leads to severe diseases and injuries along with permanent damage to specific organs of your body. You should take one gram of protein per kilogram of your body weight to maintain a healthy physique. Add vitamins and minerals in the form of fruits and vegetables to get fiber, minerals, and vitamins.

The final words on Combining the Keto Diet with Intermittent Fasting

Intermittent fasting restricts your time to eat, whereas the keto diet limits the type and amount of food. There are very few researches available on an intermittent fasting keto diet or a combination of both. So, nobody can give you a final verdict on their suitability, benefits, and risks as individual diet plans and as a combination.

But before you opt for any of these diet plans, keep it in your mind that both of them are very restrictive as compared to the other diet plans available. It may be difficult for you to stick to your eating window or to eliminate some foods from your meals for a long time. You may experience severe mood swings, stomach disorders, and low energy level at the beginning.

Have a thorough discussion with your doctor before starting intermittent fasting or keto diet. Your doctor can guide you if you opt for any one of these or a combination of both will be suitable for you. They can help you set your exercise, meals, and medicine routine according to your diet plan.

Conclusion

If you do not consider obesity as a problematic situation, you might end up piling yourself for the worst. There can be many more significant problems waiting for you. Like diabetes etc. their stages might advance as well. You would end up spending tons to your doctor. Hence it is better to control it in the initial step. You can gain control over yourself, over your mind, over your decisions and how you want your body to operate. You need to control your diet. By that, it does not mean you cannot eat what you want to. You certainly can. But you just need to know the timings that you are allowed to dine in.

Intermittent fasting is the best way to lose that bothering weight in the most natural way. Setting up a schedule is all you need, and of course, you need to follow it. Intermittent fasting has many forms and types that you can follow.

You can say goodbye to your obesity and see the difference in just a month. It is very helpful for you if you are lazy and tired of going to the gym. Stick to the plan by only being seated and watching a movie. No need for dieting obviously.

Follow your meal plan and stick to the schedule. Eat the meals that are recommended according to your time schedule. Keep a low carb diet if you want it to work effectively and quickly.

9 781802 721904